Craig Danae

UFOS UNVEILED:

A COMPREHENSIVE GUIDE TO SPOTTING

AND
UNDERSTANDING THE UNKNOWN

Craig Danae

Table of Contents

Chapter 1

Introduction to UFOs

The History of UFO Sightings

The phenomenon of unidentified flying objects (UFOs) has captivated humanity for centuries, sparking intrigue and debate across cultures and generations. Historical accounts of strange aerial phenomena can be traced back to ancient civilizations, with writings from various societies documenting lights in the sky and unusual craft. Notable early examples include the biblical account of the "wheel within a wheel" in the Book of Ezekiel and sightings recorded in ancient China. These accounts illustrate that the fascination with UFOs is not a modern invention but rather a continuation of humanity's quest to understand the unknown.

The 20th century marked a significant turning point in the history of UFO sightings, particularly following World War II. The term "flying saucer" was popularized in 1947, following Kenneth Arnold's famous sighting near Mount Rainier in Washington State, where he reported seeing nine crescent-shaped objects flying at incredible speeds. This event catalyzed a surge in UFO reports across the United States and the world, leading to government investigations and heightened public interest. The Roswell incident in 1947, where an alleged UFO crash was reported, further fueled the belief in extraterrestrial life and governmental cover-ups, making UFOs a staple of popular culture and conspiracy theories.

As the decades progressed, UFO sightings became more frequent, with notable incidents such as the 1961 Betty and Barney Hill abduction case and the 1973 Pascagoula encounter. These events not only drew media attention but also highlighted the psychological aspects of

UFO beliefs, as many witnesses reported profound transformations in their lives following their experiences. The psychological framework surrounding UFO sightings suggests that common factors, such as societal anxiety and cultural narratives, play critical roles in shaping individual perceptions of these events. Understanding these dynamics is essential for anyone interested in the deeper implications of UFO phenomena.

Advancements in technology have significantly altered the landscape of UFO sightings and reporting. With the advent of smartphones and specialized apps, enthusiasts can now document sightings more efficiently, capturing images and videos that can be analyzed for authenticity. Moreover, the rise of online communities and social media platforms has fostered collaboration among UFO believers, enabling the sharing of experiences, resources, and knowledge. Engaging with these communities can enhance your understanding of UFO sightings while providing support and validation for those who have had extraordinary encounters.

In conclusion, the history of UFO sightings is a rich tapestry woven from ancient tales, modern experiences, and technological advancements. For those looking to explore this captivating subject, it is crucial to adopt a scientific approach to analyzing sightings while remaining open to the psychological and cultural factors that influence belief. By combining effective documentation techniques, night sky observation skills, and an understanding of common misidentifications, enthusiasts can navigate the complexities of UFO sightings and contribute meaningfully to the ongoing discourse surrounding the unknown. The journey into the world of

UFOs is not just about spotting lights in the sky; it is a quest for knowledge, understanding, and perhaps even a glimpse into the extraordinary.

Cultural Impact of UFO Phenomena

Cultural beliefs surrounding UFO phenomena have evolved significantly over the decades, shaping not only societal perceptions of extraterrestrial life but also influencing various aspects of art, literature, and mass media. With the advent of the space age and the subsequent surge of UFO sightings, popular culture began to reflect a fascination with the unknown. Films, television shows, and books have depicted UFOs and alien encounters, prompting both curiosity and skepticism among audiences. This cultural engagement has served to normalize discussions surrounding unidentified flying objects, transforming them from fringe topics to mainstream interests.

The impact of UFO phenomena extends beyond entertainment. It has inspired a diverse community of enthusiasts who gather to share experiences, theories, and techniques for spotting and understanding the unknown. UFO conventions and online forums facilitate discussions that range from personal sightings to the analysis of historical cases. These gatherings not only empower individuals to explore their beliefs but also promote a sense of belonging among those who feel connected to the mysteries of the universe. As enthusiasts engage with one another, they often exchange practical knowledge on night sky observation techniques, the use of technology like smartphone apps, and advanced photography tips for capturing potential UFOs.

UFO phenomena have also influenced scientific inquiry and public policy. Historically, government bodies and military organizations have conducted investigations into UFO sightings, which has led to the establishment of more rigorous methods for documentation and reporting. This has, in turn, encouraged a more scientific approach to analyzing UFO sightings, promoting a blend of skepticism and open-mindedness. The role of weather conditions in visibility, as well as common misidentifications such as aircraft and satellites, are critical areas of study that contribute to a broader understanding of the phenomena. Such scientific discourse helps to demystify the unknown, inviting both believers and skeptics to consider the nuances behind reported sightings.

Moreover, the psychology of UFO beliefs plays a pivotal role in shaping cultural narratives surrounding these phenomena. Many individuals who report UFO encounters describe experiences that challenge their understanding of reality, leading to profound personal transformations. This psychological aspect has been explored in various academic studies, highlighting how beliefs in the supernatural can reflect deeper human desires for connection and understanding of the cosmos. As such, the cultural impact of UFO phenomena often intersects with broader existential themes, prompting individuals to question their place in the universe.

In summary, the cultural impact of UFO phenomena is multifaceted, influencing art, community, scientific inquiry, and personal psychology. As UFO believers and enthusiasts continue to explore these mysteries, they contribute to a rich tapestry of knowledge that combines personal experience with scientific rigor. Understanding the cultural implications of UFO sightings not only enhances our appreciation for the unknown but also fosters a spirit of inquiry and exploration that resonates across generations. As we delve deeper into the world of UFOs, we uncover not just the mysteries of the skies, but also the complexities of human beliefs and connections to the universe.

Purpose and Scope of the Guide

The purpose of "UFOs Unveiled: A Comprehensive Guide to Spotting and Understanding the Unknown" is to provide a thorough and accessible resource for UFO believers and enthusiasts alike. This guide aims to demystify the phenomenon of unidentified flying objects, offering practical advice, insights, and methodologies for anyone interested in exploring the skies. By addressing the spectrum of belief and experience, the guide serves as a bridge between casual observers and serious investigators, fostering a shared understanding of UFO phenomena and the curiosity that drives people to seek answers.

The scope of this guide encompasses a wide array of topics essential for both novice and experienced sky-watchers. Readers will learn how to effectively spot UFOs, including tips on night sky observation techniques that enhance visibility and improve chances of encountering unusual aerial phenomena. We delve into the common misidentifications, such as distinguishing between aircraft, satellites, and genuine UFOs, which is crucial for a balanced understanding of what one might see in the night sky. This foundational knowledge not only empowers readers but also promotes critical thinking about the nature of their sightings.

In an age defined by technology, the guide also explores how to leverage modern tools for UFO spotting. From smartphone applications that track celestial events to specialized telescopes designed for stargazing, we provide detailed recommendations on the best equipment for aspiring UFO hunters. Understanding how to utilize these technologies can greatly enhance the observation experience, making the search for the unknown both exciting and rewarding. Additionally, we address how weather conditions can affect visibility, equipping readers with the knowledge to choose the right time and place for their observations.

Photography plays a vital role in documenting UFO sightings, and this guide offers advanced tips for capturing elusive objects in the sky. Whether using a smartphone or high-end camera, readers will learn techniques for ensuring clarity and detail in their images, which is essential for any serious investigation. The guide emphasizes the importance of evidence collection, as accurate documentation can significantly contribute to the broader discourse on UFOs, helping to validate personal experiences and sightings within the community.

Finally, the guide encourages engagement with UFO communities and resources, emphasizing the value of sharing experiences and insights with like-minded individuals. We provide guidance on how to document and report sightings effectively, as well as an introduction to analyzing sightings through a scientific lens. Understanding the psychology behind UFO beliefs and sightings is also a critical component of our exploration, as it sheds light on the human experience surrounding the unknown. Together, these elements create a comprehensive framework that empowers readers to pursue their interests in UFOs with confidence and informed curiosity.

Chapter 2

How to Spot UFOs

Identifying Potential UFO Activity

Identifying potential UFO activity requires a combination of keen observation, understanding of common phenomena, and the strategic use of technology. For those who believe in the possibility of extraterrestrial visitations or have had their own unexplained experiences, developing a systematic approach to spotting UFOs can enhance the overall experience and increase the likelihood of meaningful encounters. This subchapter will explore various techniques and considerations that can aid in identifying potential UFO activity, allowing enthusiasts to refine their observational skills and engage more deeply with the unknown.

Night sky observation is fundamental to spotting UFOs. The first step is to familiarize yourself with the night sky, including the positions of stars, planets, and constellations. Utilizing star charts or mobile apps can help you learn to identify celestial bodies and enhance your observational skills. Understanding the typical patterns of commercial aircraft can also aid in distinguishing between conventional flights and potential UFOs. For instance, knowing the flight paths in your area and the characteristics of airplane lights—such as blinking and steady illumination—can help you recognize unusual movements or formations in the sky.

Weather conditions play a crucial role in UFO visibility. Clear nights with low humidity are ideal for observing the sky, as atmospheric interference can obscure objects. Furthermore, certain weather phenomena, like lenticular clouds or unusual lightning displays, can sometimes be misidentified as UFOs. Being aware of these environmental factors enables observers to interpret what they see more accurately. Additionally, certain weather conditions may spur increased UFO activity; for instance, sightings often spike during meteor showers or when unusual atmospheric conditions occur.

Incorporating technology into your UFO spotting efforts can significantly enhance your experience. Various mobile applications are available that provide real-time information about celestial events, satellite passes, and even local flight data. Telescopes and binoculars can also be valuable tools for getting a closer look at distant objects. For those interested in capturing evidence, advanced photography techniques, such as using long exposure settings and good stabilization, can yield clearer images of potential UFOs. It is essential to experiment with your equipment and understand the settings that will best capture the mysterious objects you might encounter.

Engaging with UFO communities and resources can provide additional insight and support. Online forums, local meetups, and social media groups often share sightings, tips, and resources that can enhance your UFO spotting endeavors. When documenting and reporting sightings, adopting a scientific approach can be beneficial. Keeping a detailed record of your observations, including time, location, weather conditions, and any relevant context, ensures that your experiences contribute to the broader understanding of UFO phenomena. This methodical documentation not only helps others analyze sightings but also allows for personal reflection on your encounters, deepening your connection to the mystery of the unknown.

Best Times and Locations for UFO Sightings

When it comes to spotting UFOs, timing and location play crucial roles that can significantly enhance the likelihood of an encounter. Certain times of the year, as well as specific locations, have shown a higher frequency of reported sightings, making them prime opportunities for enthusiasts to engage in their quest for the unknown. Seasonal patterns often emerge, with summer months generally providing clearer skies and longer daylight hours, perfect for both day and night observations. Additionally, during meteor showers and celestial events, the heightened activity in the night sky can sometimes coincide with increased UFO reports, making these moments ideal for those who wish to witness something extraordinary.

Geographically, some regions have become renowned for their frequent UFO sightings. Areas like the Nevada desert near Area 51, the skies over the Hudson Valley in New York, and the remote locations of the Pacific Northwest have all garnered attention from both casual observers and serious researchers. These hotspots often offer minimal light pollution and wide-open vistas, allowing for unobstructed views of the sky. For those serious about UFO spotting, aligning your location with established hotspots can increase your chances of having an encounter, as these areas tend to attract both the curious and the unexplained.

Understanding common misidentifications is essential for anyone engaging in UFO observation. Many sightings can be attributed to aircraft, satellites, weather balloons, or atmospheric phenomena. Familiarizing oneself with these common objects can significantly enhance observation skills and reduce the likelihood of misinterpreting what one sees. Utilizing technology, such as mobile apps that identify stars, satellites, and aircraft, can aid in distinguishing between ordinary objects and potential UFOs, making your observing sessions more productive. This technological assistance, combined with a keen eye, can help enthusiasts refine their spotting techniques.

Weather conditions also play a vital role in the visibility of UFOs. Clear, dry nights with low humidity are ideal for sky watching, while overcast skies, precipitation, or heavy winds can obscure the view. Wind patterns can affect the flight paths of known aircraft, leading to unusual sightings that may initially appear mysterious. Before venturing out for a night of observation, checking local weather forecasts and choosing nights with optimal conditions can greatly improve the chances of a successful sighting.

Finally, engaging with UFO communities can provide valuable insights and resources for both seasoned and novice UFO hunters. Online forums, local clubs, and social media groups often share sightings, tips, and techniques, creating a sense of camaraderie among enthusiasts. Documenting and reporting sightings effectively is crucial, as shared experiences contribute to a collective understanding of UFO phenomena. By adopting a scientific approach to analyzing sightings, enthusiasts can differentiate between genuine phenomena and misidentifications, fostering a more informed perspective on UFOs and the unexplained. The journey to uncovering the unknown is enhanced by collaboration, knowledge-sharing, and a willingness to explore the skies together.

Essential Equipment for Beginners

When venturing into the captivating world of UFO spotting, having the right equipment is crucial for beginners aiming to maximize their chances of witnessing the unexplained. Essential tools not only enhance the viewing experience but also aid in accurately documenting sightings. This section will outline the fundamental equipment every aspiring UFO enthusiast should consider, ensuring a well-prepared approach to exploring the skies.

First and foremost, a reliable pair of binoculars is an indispensable tool for any beginner. Binoculars allow users to observe distant objects with greater clarity and detail, making them perfect for scanning the night sky for anomalous lights or objects. Look for binoculars with a magnification of at least 10x and a wide field of view to ensure an optimal balance between zoom and visibility. Additionally, a sturdy tripod can enhance stability, particularly during long observation sessions. This combination of binoculars and a tripod will allow for extended viewing without the fatigue associated with holding equipment for prolonged periods.

In addition to optical equipment, utilizing technology can significantly improve the UFO spotting experience. Smartphone applications designed for stargazing can help beginners identify stars, planets, and other celestial bodies, reducing the likelihood of misidentifications. Some apps even feature augmented reality capabilities, allowing users to point their devices at the sky and receive real-time information about what they are observing. Moreover, a good quality camera equipped with a fast lens and high ISO capabilities is essential for capturing potential UFO sightings. Beginners should familiarize themselves with advanced photography techniques, such as long exposure settings and burst mode, to increase the chances of obtaining clear images of fleeting phenomena.

Weather conditions play a pivotal role in the visibility of UFOs, making a reliable weather app another critical piece of equipment. Understanding atmospheric conditions such as cloud cover, humidity, and wind patterns can significantly affect the success of spotting. Clear, dark skies are ideal for sightings, while overcast conditions may hinder visibility. Being aware of the weather forecast not only helps in planning observation nights but also prepares enthusiasts for unexpected changes that could affect their viewing experience.

Lastly, engaging with UFO communities and resources can provide valuable insights and support for beginners. Many online forums and local groups offer equipment recommendations, sharing experiences and techniques that can enhance personal sightings. These communities often host events where enthusiasts can gather, share their equipment, and learn from one another. By documenting and reporting sightings effectively, beginners can contribute to a larger body of knowledge that aids in the scientific analysis of UFO phenomena. This collaboration not only enriches the individual experience but also fosters a sense of connection among UFO enthusiasts, promoting a culture of shared discovery and understanding in the pursuit of the unknown.

Chapter 3

Night Sky Observation Techniques

Understanding the Night Sky

Understanding the night sky is a crucial first step for anyone interested in spotting unidentified flying objects (UFOs). The vast expanse above us is filled with celestial bodies, atmospheric phenomena, and human-made objects that can easily be mistaken for something otherworldly. To differentiate between the familiar and the truly unidentified, one must develop a keen observational skill set. This involves not only knowing how to identify stars and constellations but also being aware of the various types of aircraft, satellites, and atmospheric conditions that can affect visibility. The night sky can be both enchanting and perplexing, making it essential to approach it with a discerning eye.

Night sky observation techniques play a pivotal role in enhancing your ability to spot UFOs. Begin by familiarizing yourself with the major constellations and their seasonal shifts. This foundational knowledge will help you navigate the sky more effectively. Additionally, understanding the movement of satellites, which are often visible as fast-moving points of light, can help eliminate common misidentifications. Techniques such as maintaining a consistent observation schedule and using star charts or apps can significantly improve your chances of spotting something unusual. The more time you spend under the stars, the more you'll begin to recognize what is typical and what may warrant further investigation.

Misidentifications are one of the biggest hurdles in UFO spotting. Many enthusiasts have reported seeing strange lights or objects in the sky, only to later find out they were observing conventional aircraft or satellites. For instance, the International Space Station (ISS) is particularly bright and can be mistaken for a UFO due to its swift passage across the sky. Understanding these common misidentifications allows enthusiasts to approach sightings with a more analytical mindset. Educating yourself on the characteristics of various celestial bodies and human-made objects can be the key to distinguishing the extraordinary from the mundane.

Advancements in technology have greatly enhanced our ability to spot and document UFOs. Today, several apps are available that can help you identify stars, planets, and satellites in real-time, providing valuable context during your observations. Telescopes equipped with night vision capabilities can also open up new avenues for exploration. However, while technology can aid in identification, it is essential to combine these tools with fundamental observational skills to achieve the best results. The integration of technology into your observational practices can elevate your understanding of the night sky, making it a more rewarding experience.

Lastly, weather conditions significantly impact UFO visibility. Clear skies provide the best opportunities for observation, while cloud cover, humidity, and atmospheric turbulence can obscure celestial objects. Being aware of local weather patterns and using weather apps can help you plan your observation sessions more effectively. Engaging with UFO communities can also provide insights into the best times and locations for sightings, as well as shared experiences and techniques. By documenting and reporting your sightings, you contribute to a collective understanding of the phenomenon, enhancing both personal and communal knowledge of the unknown.

Star Gazing vs. UFO Spotting

In exploring the fascinating realm of unidentified flying objects (UFOs), it is essential to distinguish between the art of star gazing and the pursuit of UFO spotting. While both activities involve observing celestial phenomena, their objectives and techniques diverge significantly. Star gazing primarily focuses on appreciating the beauty of the night sky, identifying constellations, and understanding astronomical bodies such as stars, planets, and satellites. In contrast, UFO spotting is driven by the desire to encounter and document unexplained aerial phenomena, often with the intent of uncovering evidence of extraterrestrial life or advanced technology. This subchapter delves into the nuances of these two practices, providing insights into how enthusiasts can enhance their experiences in both fields.

To effectively spot UFOs, one must first develop a keen understanding of the night sky and the various elements within it. Familiarity with common celestial objects is crucial, as misidentification can lead to false claims of UFO sightings. For instance, knowing the difference between a satellite, an aircraft, or a natural astronomical phenomenon like a meteor shower can prevent confusion and enhance credibility when reporting sightings. Engaging in star gazing as a foundational practice can significantly improve one's observational skills, allowing enthusiasts to recognize anomalous behavior in the sky that may warrant further investigation.

Weather conditions play a pivotal role in both star gazing and UFO spotting, influencing visibility and the likelihood of experiencing extraordinary events. Clear nights with minimal light pollution are ideal for both activities, but specific atmospheric phenomena can enhance or obscure visibility. For instance, certain cloud formations may create optical illusions that could be mistaken for UFOs, while atmospheric refraction can distort the appearance of stars and planets. Understanding how weather impacts these observations enables enthusiasts to choose the best nights for their pursuits and enhances their ability to discern genuine UFO sightings from ordinary atmospheric events.

Modern technology has revolutionized the way we engage with both star gazing and UFO spotting. A variety of apps are available that help users identify stars, planets, and satellites in real-time, making it easier to track celestial movements and recognize potential UFO activity. Additionally, advanced photography techniques can aid in capturing evidence of UFOs, helping enthusiasts document their sightings more effectively. Investing in quality telescopes and cameras can enhance the likelihood of capturing clear images, which are invaluable for analysis and reporting to UFO communities. These technological tools not only facilitate personal exploration but also contribute to a broader understanding of aerial phenomena.

Lastly, the psychological aspect of UFO beliefs is a crucial consideration in this discussion. Many individuals who engage in UFO spotting often come from backgrounds of mysterious experiences or supernatural encounters. The communities built around these interests foster a sense of belonging and shared curiosity, providing a platform for reporting sightings and discussing theories. Documenting and analyzing sightings within these communities can lead to a more scientific approach to understanding the unknown, which ultimately bridges the gap between star gazing and UFO spotting. By fostering an open dialogue and encouraging rigorous investigation, enthusiasts can contribute to a more nuanced understanding of both the cosmos and the potentially extraordinary phenomena that inhabit it.

Using Binoculars and Telescopes

The exploration of unidentified flying objects (UFOs) can be significantly enhanced through the use of binoculars and telescopes. These tools not only provide a closer view of the night sky but also help enthusiasts distinguish between common aerial phenomena and potential extraterrestrial encounters. Understanding the features and functionalities of these optical instruments is essential for anyone interested in expanding their observational skills and increasing their chances of spotting UFOs.

When selecting binoculars for UFO observation, it is crucial to consider the specifications that best suit your needs. Binoculars with a higher magnification and larger objective lens diameter are ideal for night sky observation, as they allow for greater light collection and clearer images of distant objects. A pair with an 8x42 magnification, for instance, strikes a balance between portability and powerful viewing capabilities. It's also beneficial to choose models with a wide field of view, which can help track fast-moving objects that may not be easily visible with the naked eye.

Telescopes offer an even more detailed perspective, making them invaluable for serious UFO enthusiasts. A good telescope can reveal intricate details of celestial bodies, allowing for the identification of satellites, planets, and other astronomical phenomena. When using a telescope for UFO spotting, it is important to set it up in a location with minimal light pollution and stable atmospheric conditions. Additionally, using a computerized mount can assist in tracking objects more easily, especially those that appear suddenly and move quickly across the sky.

However, it is essential to remain mindful of common misidentifications when using these tools. Aircraft, weather balloons, and satellites can all appear similar to UFOs under certain conditions. Understanding the characteristics of these objects, such as their flight patterns and typical brightness, can help in distinguishing them from genuine unidentified phenomena. Utilizing apps that provide real-time information about air traffic, satellite timings, and astronomical events can further enhance your observational accuracy and reduce the likelihood of misinterpretation.

The effectiveness of using binoculars and telescopes for UFO spotting is also influenced by weather conditions. Clear skies with minimal atmospheric turbulence provide the best visibility, while humidity, clouds, and light pollution can significantly obstruct views. It is advisable to check local weather forecasts and choose nights with optimal viewing conditions for your observations. Combining these tools with proper observation techniques and a keen eye for detail will not only improve your chances of spotting a UFO but also deepen your understanding of the night sky, enriching your overall experience in the pursuit of the unknown.

Chapter 4

Understanding Common Misidentifications

Aircraft and Drones

In the realm of unidentified flying objects (UFOs), distinguishing between aircraft, drones, and genuine anomalies can be a challenging yet essential skill for enthusiasts. Understanding the characteristics of conventional flying machines and the emerging technology of drones is crucial for anyone interested in spotting UFOs. Aircraft, including commercial airliners, private planes, and military jets, have specific flight patterns, altitudes, and operational features that can often lead to misidentifications. Familiarity with these elements will empower observers to discern what they are witnessing in the night sky.

Drones have become increasingly prevalent in both urban and rural landscapes, adding another layer of complexity to UFO sightings. These unmanned aerial vehicles can operate at varying altitudes, perform agile maneuvers, and emit lights that may appear unusual to the untrained eye. When attempting to identify a potential UFO, it is important to consider the possibility that what you are seeing might be a drone, especially if it exhibits erratic flight patterns or lights that change in intensity or color. By understanding the capabilities and limitations of drones, observers can refine their skills in distinguishing them from more enigmatic phenomena.

The role of technology in UFO spotting cannot be overstated. Various mobile applications and devices have been developed to assist enthusiasts in tracking aerial objects and understanding their nature. For example, flight tracking apps can reveal the location and flight path of nearby aircraft, while astronomy tools help users identify celestial objects such as satellites or planets. The use of telescopes and binoculars can enhance night sky observation, allowing for closer examination of objects that may initially appear to be unidentified. Incorporating these technological aids into your UFO spotting endeavors can greatly enhance your observational accuracy.

Weather conditions significantly impact the visibility of aerial phenomena. Clear skies with minimal atmospheric interference provide the best opportunities for spotting UFOs, while fog, rain, or overcast conditions can obscure the view. Additionally, phenomena such as temperature inversions can create illusions that may lead to misidentifications. Understanding how weather interacts with aerial visibility will equip observers with the knowledge necessary to optimize their UFO watching sessions. It's beneficial to keep a weather journal, noting conditions during sightings, as this can help in analyzing patterns or potential correlations.

Lastly, effective documentation and reporting of UFO sightings are paramount for advancing our understanding of these events. Keeping detailed records of sightings, including time, location, weather conditions, and any technological aids used, can contribute to a broader database of UFO encounters. Engaging with UFO communities and sharing experiences not only fosters a supportive environment for enthusiasts but also aids in collective analysis. By employing a scientific approach to studying sightings, individuals can transform personal experiences into valuable contributions to the ongoing discourse surrounding UFOs, bridging the gap between belief and empirical observation.

Satellites and Space Debris

Satellites play a crucial role in our understanding of the night sky and the phenomena that inhabit it, including UFO sightings. Launched by various governments and private entities, these objects orbit our planet for purposes ranging from communication to weather forecasting. For those interested in spotting UFOs, recognizing the characteristics of satellites can significantly reduce misidentification. Satellites often travel in predictable paths and can be seen reflecting sunlight, creating bright points of light that move steadily across the sky. By familiarizing yourself with satellite schedules and trajectories, you can better distinguish them from potential UFOs.

The increasing number of satellites in low Earth orbit has raised concerns about space debris. As more satellites are launched, the likelihood of collisions and the creation of debris fields grows. This debris can also appear in the night sky as fast-moving streaks or flashes, leading to further confusion for UFO observers. It is essential to be aware of this phenomenon, as space debris can mimic the characteristics typically associated with unidentified flying objects. Understanding the distinction between operational satellites and debris can enhance your ability to analyze what you see in the sky.

For those keen on utilizing technology for spotting UFOs, various apps and websites can provide real-time tracking of satellite movements. Tools like Heavens-Above and Star Walk allow users to map out the night sky, showing when and where satellites will be visible. This technological engagement not only cultivates a deeper understanding of celestial mechanics but also empowers enthusiasts to document their sightings more accurately. By integrating satellite tracking into your night sky observation routine, you can refine your skills in distinguishing between known and unknown objects.

Weather conditions play a vital role in visibility, impacting both satellite observation and the potential for UFO sightings. Clear skies are ideal for observing satellites, as clouds and atmospheric disturbances can obscure your view. Additionally, various weather phenomena, such as auroras or lightning, may contribute to misinterpretations of what is seen. Understanding how these conditions affect visibility can help you keep a clear perspective on what you are observing, ensuring that your excitement about potential UFOs is not muddied by environmental variables.

Lastly, engaging with UFO communities can enhance your knowledge about spotting techniques and the nuances of distinguishing between satellites, space debris, and actual unidentified flying objects. Online forums, social media groups, and local meet-ups provide platforms for sharing experiences, tips, and insights. Documenting sightings with precision and reporting them effectively can also contribute to a collective understanding of the phenomena. By collaborating with fellow enthusiasts and utilizing technology to track satellites and space debris, you will not only improve your observational skills but also enrich the broader discourse surrounding UFOs and the mysteries they present.

Natural Phenomena

Natural phenomena play a crucial role in the study of UFO sightings, often acting as both a backdrop and a source of confusion for observers. Understanding these phenomena is essential for anyone interested in distinguishing between the extraordinary and the mundane. Many UFO sightings can be attributed to natural events, such as atmospheric anomalies, celestial bodies, or meteorological conditions. By becoming familiar with these natural occurrences, UFO enthusiasts can sharpen their observational skills and enhance their ability to identify true unidentified flying objects.

One of the most common natural phenomena mistaken for UFOs is atmospheric refraction. This phenomenon occurs when light bends as it passes through layers of air with varying temperatures and densities, creating optical illusions. For instance, a distant aircraft may appear distorted or take on unusual shapes due to this bending effect. Additionally, light from the setting sun can create spectacular displays, such as sun dogs or halos, which can easily be misidentified as alien spacecraft. Learning to recognize these effects is vital for effective night sky observation, allowing enthusiasts to filter out false positives from their sightings.

Celestial bodies, including planets, stars, and meteor showers, frequently contribute to UFO reports. Bright planets like Venus or Jupiter can shine intensely and appear to move across the sky, particularly when they are low on the horizon. Similarly, meteor showers can produce fleeting bright streaks that may be misconstrued as UFOs. By studying the patterns of celestial movements and familiarizing oneself with astronomical events, observers can significantly reduce the likelihood of misidentification. Utilizing technology, such as astronomy apps, can aid in tracking these events and improving observational accuracy.

Weather conditions also play a pivotal role in the visibility of potential UFOs. Atmospheric phenomena such as clouds, fog, and storms can obscure views of the night sky or create unusual lighting effects. For example, lenticular clouds can form in a way that resembles a flying saucer, while lightning can create transient flashes that might be mistaken for advanced aircraft. Understanding how weather influences visibility can help enthusiasts choose the best times and conditions for their skywatching activities, enhancing their chances of spotting genuine unidentified objects.

Finally, the psychological aspect of UFO sightings cannot be overlooked. Many individuals who report UFO encounters describe feelings of awe or fear, which can influence their perceptions. This psychological component is intertwined with natural phenomena, as the unusual and unexplainable can evoke strong emotional responses. Engaging with UFO communities and resources can provide a supportive environment for sharing experiences and insights, fostering a deeper understanding of both the natural world and the psychological factors that shape our beliefs. By cultivating a balanced perspective, enthusiasts can approach their skywatching endeavors with both excitement and critical thinking, ultimately enriching their quest for the truth behind the unknown.

Chapter 5

Using Technology for UFO Spotting

Recommended Apps for Sky Observation

In the ever-evolving landscape of UFO research and observation, technological advancements have made it easier than ever for enthusiasts to engage with the night sky. Numerous applications are available that enhance the experience of sky observation, allowing users to identify celestial objects and spot potential unidentified flying objects. This subchapter explores some of the most recommended apps for sky observation, tailored specifically for those intrigued by the supernatural and the potential for extraterrestrial encounters.

One of the leading apps for night sky enthusiasts is SkySafari. This powerful tool offers a comprehensive database of stars, planets, and other celestial bodies, providing users with detailed information and visual representations of what they are observing. SkySafari's augmented reality feature allows users to point their device at the sky and see a real-time overlay of celestial objects, making it easier to distinguish between planets, stars, and potential UFOs. Its extensive catalog also includes historical data, which can help users understand past sightings and the possible origins of unidentified objects.

Another excellent option is Stellarium, a free and open-source planetarium app designed to simulate the night sky in real-time. This app is user-friendly and provides accurate representations of the night sky from any location on Earth. With its ability to show satellite paths and the locations of known aircraft, Stellarium helps users differentiate between common aerial phenomena and potential UFOs. The app's night mode feature is particularly useful for preserving night vision, allowing for a more immersive observation experience.

For those interested in documenting their sightings, the UFO Detection app is specifically designed for UFO enthusiasts. This app allows users to log their sightings, upload photographs, and share their experiences with a community of like-minded individuals. It offers a unique platform for analysis, where users can compare their sightings with others and gain insights into common patterns and characteristics of UFO phenomena. The app also includes a reporting feature, which guides users on how to document their sightings effectively, ensuring that crucial details are captured for future reference.

Weather conditions play a significant role in sky observation, and apps like Weather Underground provide real-time updates on atmospheric conditions. Understanding the weather is essential for optimal UFO spotting, as factors such as cloud cover, humidity, and light pollution can affect visibility. By integrating weather data with the use of astronomical apps, observers can plan their skywatching sessions more effectively, increasing the likelihood of encountering something extraordinary. Combining these resources creates a powerful toolkit for any UFO believer eager to explore the mysteries of the night sky.

In conclusion, the integration of technology into sky observation has transformed how enthusiasts engage with the universe. By utilizing apps like SkySafari, Stellarium, UFO Detection, and Weather Underground, individuals can enhance their understanding of celestial phenomena and improve their chances of spotting UFOs.

As the boundaries between science and the supernatural continue to blur, these tools provide essential support for those seeking to document, analyze, and comprehend the unknown that lies above us.

Utilizing Drones for Enhanced Viewing

Utilizing drones for enhanced viewing presents an innovative approach for those deeply invested in the study of UFO phenomena. As technology continues to evolve, drones have emerged as powerful tools for aerial observation, offering unique perspectives that were previously unattainable. These unmanned aerial vehicles (UAVs) can cover vast areas quickly, providing real-time footage and data that could significantly aid in identifying unusual aerial phenomena. For UFO enthusiasts, this means not only increasing the chances of spotting a UFO but also enhancing the overall experience of sky observation.

One of the primary advantages of using drones is their ability to reach altitudes and locations that might be difficult for human observers. Equipped with high-resolution cameras and advanced stabilization technology, drones can capture clear images and videos of the night sky and surrounding areas. This capability allows users to monitor potential UFO hotspots from above, increasing the likelihood of encountering unidentified flying objects. Moreover, drones can be programmed to follow specific flight paths, allowing for systematic exploration of regions that have a history of UFO sightings or other supernatural activities.

In addition to their observational capabilities, drones can be outfitted with various sensors that enhance the detection of unusual phenomena. For instance, thermal imaging cameras can help identify objects that emit heat, which may be useful in distinguishing between natural occurrences and man-made objects. Similarly, drones with spectrometers can analyze light emissions, potentially revealing anomalies that are not visible to the naked eye. By leveraging these advanced technologies, UFO enthusiasts can gather more comprehensive data, improving their ability to document and analyze sightings effectively.

Weather conditions play a crucial role in visibility, both for drones and for spotting UFOs. Understanding how atmospheric factors like cloud cover, humidity, and light pollution can impact drone operations is vital for maximizing their effectiveness in UFO investigations. For optimal results, enthusiasts should consider launching drones during clear nights, away from urban light sources, to ensure the best possible observational conditions. Additionally, being aware of local weather patterns can help in planning expeditions, allowing for timely responses to potential sightings as conditions change.

Finally, engaging with UFO communities and resources can amplify the benefits of utilizing drones for enhanced viewing. By sharing experiences, insights, and findings with fellow enthusiasts, individuals can foster a collaborative environment that promotes further research and understanding of UFO phenomena. Online forums, social media groups, and local meet-ups can serve as valuable platforms for exchanging tips on drone usage, discussing sighting reports, and exploring the latest advancements in technology. This collective knowledge can empower individuals to become more adept at spotting UFOs, ultimately contributing to the broader quest for understanding the mysterious and the unexplained.

The Role of Night Vision Equipment

The role of night vision equipment in the quest to uncover the mysteries of unidentified flying objects (UFOs) cannot be overstated. As enthusiasts and believers gather under the cloak of night to scan the skies, having the right tools can significantly enhance their observational capabilities. Night vision technology, which amplifies available light to allow for visibility in low-light conditions, offers a unique advantage in spotting potential UFOs. These devices range from binoculars and monoculars to more advanced goggles, each serving to illuminate the darkness and reveal phenomena that the naked eye might miss.

Among the various types of night vision equipment, digital night vision devices have gained popularity for their ability to provide real-time video feed and capture imagery. Unlike traditional night vision equipment, which relies on intensifying ambient light, digital models can record and store sightings, enabling enthusiasts to review and analyze their observations later. This capability is particularly beneficial for documenting UFO sightings, as it allows for a more thorough examination of the captured footage. Additionally, many of these devices come equipped with features like infrared capabilities, which can further enhance visibility during nighttime excursions.

While night vision equipment can significantly improve the chances of spotting a UFO, it is essential to understand the limitations and considerations associated with its use. Factors such as weather conditions, moonlight, and atmospheric clarity can impact visibility.

For instance, fog or heavy cloud cover can obscure the view, making it difficult to discern objects in the sky, regardless of the quality of the night vision gear. Enthusiasts should also be aware of the potential for common misidentifications, such as aircraft or satellites, which can appear unusual when viewed through night vision optics. A well-rounded understanding of these elements is crucial for effective sky observation.

Engaging with UFO communities can also enhance the experience of using night vision equipment. Sharing insights and experiences with like-minded individuals can lead to valuable tips and techniques for maximizing the effectiveness of night vision during skywatching sessions. Many enthusiasts recommend joining local groups or online forums where members discuss their sightings, equipment preferences, and the best locations for observing the night sky. This collaborative approach not only fosters a sense of community but also enriches the collective knowledge regarding UFO sightings and the technologies used to spot them.

Ultimately, the integration of night vision equipment into the UFO spotting experience represents a fusion of technology and curiosity. With the right tools and a thoughtful approach, enthusiasts can elevate their observational skills, making the night sky more accessible and revealing. As they venture out into the darkness, equipped with these advanced technologies, they stand a better chance of uncovering the secrets that lie overhead, while simultaneously contributing to a broader understanding of the phenomena that continue to captivate the human imagination.

Chapter 6

The Role of Weather Conditions in UFO Visibility

How Weather Affects Sightings

Weather plays a crucial role in the visibility and frequency of UFO sightings, impacting not only the clarity of the sky but also the conditions under which observers are likely to be looking up. Atmospheric conditions such as cloud cover, humidity, and precipitation can either obscure or enhance the visibility of aerial phenomena. For instance, a clear night with minimal light pollution allows for optimal viewing conditions, making it easier to spot objects that might otherwise go unnoticed. Conversely, fog, rain, or heavy cloud cover can render the night sky virtually invisible, drastically reducing the chances of encountering unidentified flying objects.

Temperature and atmospheric pressure also play significant roles in how phenomena are perceived by observers. Changes in temperature can affect the refraction of light, potentially distorting the appearance of distant objects. This distortion can lead to misidentification of terrestrial objects such as aircraft or satellites as UFOs. Moreover, certain atmospheric phenomena, like temperature inversions, can create illusions that might confuse even seasoned skywatchers, making it essential to consider weather conditions when analyzing sightings.

Wind patterns can further complicate the situation. High winds can create turbulence that affects the stability of flying objects, potentially making them appear erratic or unusual in their movements. Additionally, weather-generated phenomena such as ball lightning or lenticular clouds can be mistaken for UFOs due to their unconventional shapes and behaviors.

Understanding these meteorological factors can aid enthusiasts and researchers alike in distinguishing between genuine anomalous sightings and natural occurrences.

The use of technology, including apps and telescopes, can also be influenced by weather conditions. Many modern UFO spotting applications provide real-time data about satellite movements and aircraft activity, which can help users eliminate common misidentifications. However, these technologies may become less effective in adverse weather conditions where visibility is compromised. Photographic techniques, too, must be adjusted according to the weather; for instance, capturing clear images of UFOs requires optimal lighting and atmospheric clarity, which are often not present during stormy or overcast nights.

Ultimately, understanding the intricate relationship between weather conditions and UFO sightings is essential for those dedicated to the pursuit of the unknown. By considering meteorological factors, enthusiasts can refine their observation strategies, enhance their documentation efforts, and contribute more effectively to the broader discourse on unidentified aerial phenomena. This knowledge not only cultivates a deeper appreciation for the complexities of skywatching but also empowers individuals to approach their experiences with a more informed and analytical perspective.

Seasonal Considerations

Seasonal considerations play a crucial role in the pursuit of UFO sightings and understanding the phenomena associated with them. Each season brings unique environmental factors that can either enhance or hinder your ability to spot unidentified flying objects. Changes in weather patterns, atmospheric conditions, and even the position of celestial bodies can significantly influence visibility. As enthusiasts and believers, it is essential to be aware of these seasonal variations to maximize your chances of encountering the unknown.

Spring and summer months often provide the clearest skies, making them prime seasons for night sky observation. The longer daylight hours and warmer temperatures encourage more people to engage with the outdoors, leading to increased reports of sightings. During these seasons, the prevalence of atmospheric disturbances tends to decrease, resulting in clearer skies that facilitate the observation of celestial events such as meteor showers and planetary alignments. However, it is also essential to remain vigilant about common misidentifications, as many individuals may mistake aircraft or satellites for UFOs during these active months.

Autumn introduces a transition in the atmosphere that can affect visibility. As temperatures begin to drop, air quality may improve, and the likelihood of atmospheric turbulence decreases, often leading to clearer nights. This is an excellent time for advanced photography, as the crisp air can enhance the definition of images captured in the night sky. However, the changing foliage can also obstruct views, so finding open spaces becomes critical.

Additionally, the onset of winter can bring about its own challenges. Snowy landscapes reflect light, which can create visual illusions, and inclement weather often limits sky observation opportunities.

Utilizing technology becomes even more essential during the winter months. Apps designed for stargazing can help track celestial movements and identify potential UFOs among the stars. Telescopes may also be beneficial, but they require careful setup and patience in cold conditions. Engaging with UFO communities can provide valuable insights on seasonal trends and sightings, facilitating a shared knowledge base that can further enhance your observational skills. By collaborating with fellow enthusiasts, you can exchange tips about the best locations and techniques suited for the season at hand.

Documenting and reporting sightings effectively is vital, especially when seasonal variations impact the nature of UFO phenomena. Each season may bring different types of sightings, influenced by both human activity and natural occurrences. A scientific approach to analyzing these sightings can reveal patterns that enhance our understanding of UFO behavior. Furthermore, considering the psychology behind seasonal beliefs in the supernatural can shed light on why certain times of the year may foster increased sightings or heightened interest in UFOs. By integrating all these seasonal considerations into your UFO spotting strategies, you will be better equipped to navigate the mysteries of the sky and contribute meaningfully to the ongoing discourse surrounding unidentified aerial phenomena.

Best Practices for Different Weather Scenarios

When it comes to observing UFOs, understanding how different weather scenarios can impact visibility and detection is crucial. Conditions such as cloud cover, precipitation, and atmospheric clarity can either enhance or hinder the chances of spotting unidentified aerial phenomena. UFO enthusiasts should familiarize themselves with these variables to maximize their observational efforts. This subchapter outlines best practices tailored to various weather conditions, ensuring that aspiring spotters are well-prepared regardless of what the skies present.

On clear nights, when stars shine brightly and atmospheric conditions are stable, the chances of spotting a UFO increase significantly. It's advisable to find a location away from light pollution, ideally in rural areas or designated dark sky parks. Use a star chart or an astronomy app to orient yourself with the night sky, which can help in distinguishing between celestial objects and potential UFOs. During such nights, employing advanced photography techniques—like long exposure settings—can aid in capturing elusive objects. It's also beneficial to bring binoculars or a telescope to get a closer look at distant phenomena, while remembering to keep an eye out for sudden movements that could indicate something out of the ordinary.

Conversely, cloudy or overcast conditions can pose challenges for UFO spotting. In such instances, utilizing technology becomes essential. Infrared cameras or night vision devices can penetrate low-light conditions and provide a clearer view of the sky. These tools can help detect heat signatures or movements that are otherwise invisible to the naked eye. Furthermore, keeping track of weather forecasts can help in planning expeditions, as isolated clear patches or breaks in the clouds may present unexpected opportunities for sightings. Engaging with local meteorological reports can also help enthusiasts understand the patterns that might lead to enhanced visibility.

Rainy or stormy weather tends to obscure visibility, but it also introduces unique opportunities for spotting UFOs. Phenomena such as lightning can sometimes coincide with unidentified aerial sightings, leading to unusual visual displays. Observers should remain cautious and prioritize safety, but utilizing waterproof gear and equipment can facilitate continued observation during light showers. Additionally, documenting any unusual occurrences during inclement weather can lead to interesting insights and patterns in UFO sightings, as some enthusiasts believe that certain weather conditions may attract or influence these phenomena.

In extreme weather conditions, such as heavy fog or snow, patience and adaptability become key. Visibility might be severely limited, yet this does not mean that sightings are impossible. Engaging with local UFO communities can provide insight into historical sightings that occurred during similar conditions, and sharing experiences can lead to collective understanding. Effective documentation plays a critical role during these times; enthusiasts should take notes on any unusual sounds, lights, or movements they perceive, as these details can be crucial for later analysis. By being prepared and adapting to the weather, UFO enthusiasts can enhance their chances of making meaningful observations, regardless of the challenges presented by the elements.

Chapter 7

Advanced Photography Tips for Capturing UFOs

Choosing the Right Camera and Lens

Choosing the right camera and lens is crucial for anyone serious about documenting UFO sightings. A well-equipped observer can significantly increase their chances of capturing evidence of the unknown, and understanding the technical aspects can make all the difference. The first consideration should be the type of camera. DSLR and mirrorless cameras are often recommended for their versatility, image quality, and the ability to interchange lenses. These cameras allow for manual settings, which are essential for night sky photography, where light conditions can be challenging. Additionally, they offer the option to shoot in RAW format, preserving more detail for post-processing.

When selecting a lens, the focal length plays a pivotal role in UFO photography. A wide-angle lens can be beneficial for capturing large swathes of the sky, making it easier to identify unusual objects. However, a telephoto lens is equally important for zooming in on distant objects, which is often necessary when attempting to document UFOs. A lens with a fast aperture (e.g., f/2.8 or lower) will allow more light to enter the camera, which is particularly advantageous during night observations. This is because many UFO sightings occur under low-light conditions, and a lens with a wider aperture can help capture clearer images.

Stability is another critical factor in ensuring successful documentation. A sturdy tripod is recommended, as it provides stability while reducing camera shake, especially during long exposure shots. Long exposures are often required to gather enough light in dark conditions, making a tripod an invaluable tool. Additionally, using a remote shutter release can further minimize shake, allowing for sharper images. For those who may not have access to high-end equipment, even a smartphone with a good camera can be used effectively, especially when combined with stabilization techniques and specialized apps designed for astrophotography.

Weather conditions can also influence visibility and the effectiveness of your equipment. Clear skies are ideal for UFO spotting, but it is essential to consider how different weather elements can affect both visibility and camera performance. For instance, humidity can cause lens fogging, while high wind can shake your setup. Being aware of the weather forecast and planning your observation sessions accordingly can greatly enhance your chances of success. Additionally, using apps that track satellite movements and astronomical events can help you distinguish between natural phenomena and potential UFOs.

Finally, engaging with UFO communities can provide insights into the best practices for capturing and documenting sightings. Online forums and local groups often share tips and experiences related to equipment and techniques. By interacting with fellow enthusiasts, you can learn about the latest technology, share your findings, and even collaborate on observation nights. The collective knowledge of the community can guide you toward making informed decisions regarding your photography setup, ultimately enhancing your ability to document the unexplained effectively.

Techniques for Long Exposure Photography

Long exposure photography is an essential technique for UFO enthusiasts looking to capture evidence of unidentified aerial phenomena against the backdrop of the night sky. This method allows for the accumulation of light over an extended period, which can reveal details that might be invisible to the naked eye. By utilizing long exposure techniques, you can enhance your chances of documenting UFOs, stars, and other celestial objects, contributing valuable data to the ongoing discourse surrounding extraterrestrial observations.

To successfully implement long exposure photography, it is vital to understand the equipment needed. A sturdy tripod is essential to prevent any camera shake during the exposure. DSLRs or mirrorless cameras equipped with manual settings are preferred for this purpose, as they allow for adjustments in shutter speed, ISO, and aperture. Additionally, using a remote shutter release can further minimize vibrations, ensuring that the images captured are as clear as possible. If you are using a smartphone, various apps simulate long exposure effects, although they may not offer the same quality as dedicated cameras.

When setting up for a long exposure shot, choosing the right location is crucial. Ideally, you should be away from urban light pollution, which can obscure faint objects in the sky. Dark sites, such as rural areas or designated dark sky parks, provide optimal conditions for night sky photography. It is also important to consider the weather conditions; clear skies without significant cloud cover are ideal for visibility. Check for moon phases as well, as a bright moon can wash out the night sky, making it difficult to spot and capture UFOs or other celestial phenomena.

Once you have found the perfect spot and set up your equipment, it's time to adjust your camera settings. A common starting point for long exposure photography is a shutter speed of 15 to 30 seconds, with an ISO setting between 800 and 3200, depending on the ambient light. Aperture should generally be set to a wider opening (lower f-stop number) to allow more light into the camera. Experimenting with these settings can yield different results; therefore, it's beneficial to take multiple shots to find the optimal configuration for your specific environment and lighting conditions.

After capturing your images, the post-processing stage is where the magic happens. Utilizing photo editing software can enhance your long exposure shots, revealing details that may have been missed during the initial capture. Techniques such as adjusting contrast, brightness, and saturation can bring out the features of the UFOs or celestial objects present in your photographs. Additionally, sharing your findings with UFO communities online can provide valuable feedback and further insights, helping you connect with fellow enthusiasts and contribute to a broader understanding of the phenomena you are documenting. By mastering the techniques of long exposure photography, you can significantly enhance your ability to record and analyze potential UFO sightings.

Post-Processing Tips for UFO Images

Post-processing is an essential step in enhancing UFO images, allowing enthusiasts and researchers to extract as much detail and clarity as possible from their captures. After a sighting, many individuals rush to share their images online or present them to communities, but taking the time to refine and analyze these images can significantly impact the credibility and comprehensibility of the evidence. This section will guide you through effective post-processing techniques that can improve your UFO photographs, ensuring they convey the mystery and intrigue of the unknown while remaining grounded in clear, visual analysis.

One of the primary tools in post-processing is image editing software. Programs like Adobe Photoshop or GIMP offer a range of features that can be utilized to adjust brightness, contrast, and sharpness. Enhancing these elements can help bring out details that may not be visible in the original photograph. For instance, increasing contrast can help delineate the edges of a UFO against the night sky, making it easier to identify shapes and structures that might otherwise blend into the background. When utilizing these tools, it is crucial to maintain the integrity of the image, avoiding over-editing that could misrepresent the sighting.

Another vital aspect of post-processing is noise reduction. Many UFO images are captured in low-light conditions, which can introduce graininess or pixelation. Software tools that specialize in noise reduction can smooth out these imperfections without sacrificing detail, allowing for a clearer view of the object. Additionally, applying filters strategically can help highlight features of the UFO while minimizing distractions from the surrounding environment. However, enthusiasts should always be cautious, as excessive filtering can lead to artifacts that may raise skepticism about the authenticity of the image.

Incorporating metadata analysis into your post-processing routine is another important consideration. Metadata can provide valuable context about the conditions under which the photograph was taken, such as time, location, and camera settings. This information can help corroborate the sighting and add credibility to your documentation. By combining visual analysis with metadata, you can build a more comprehensive narrative around your experience, making it easier for others to understand the significance of the sighting.

Lastly, sharing your processed images within UFO communities can foster discussion and collaboration. Engaging with fellow enthusiasts allows for a collective analysis of sightings, where various perspectives can lead to new insights. Be open to constructive feedback and consider how your post-processed images can contribute to ongoing dialogues about unidentified aerial phenomena. By documenting and presenting your findings thoughtfully, you not only enhance your own understanding but also contribute to the larger discourse surrounding UFO phenomena.

Chapter 8

Engaging with UFO Communities and Resources

Online Forums and Social Media Groups

Online forums and social media groups have emerged as vital platforms for UFO enthusiasts, allowing individuals to connect, share experiences, and gather insights on the mysteries of the skies. These digital communities provide a space for believers and skeptics alike to discuss sightings, theories, and technologies related to unidentified flying objects. By participating in these online spaces, individuals can access a wealth of information, broaden their perspectives, and build connections with like-minded people who share a fascination for the unknown.

One of the primary benefits of online forums and social media groups is the ability to exchange personal experiences and firsthand accounts. Members often recount their sightings, detailing the conditions and contexts in which they occurred. Such narratives can be invaluable for those learning how to spot UFOs or for those trying to refine their night sky observation techniques. By aggregating these accounts, aspiring UFO spotters can better understand what to look for and how to differentiate between genuine phenomena and common misidentifications, such as commercial aircraft or satellites.

Technology plays a significant role in facilitating these discussions, with many groups embracing apps and devices that aid in UFO spotting. Members frequently share recommendations for advanced photography techniques, which can help capture fleeting moments in the sky. These shared resources often include tutorials on using telescopes or smartphone applications that track celestial events. By leveraging these tools, enthusiasts can enhance their observational skills, improving their chances of documenting potential UFO sightings.

Weather conditions significantly impact the visibility of UFOs, and many online forums dedicate threads to discussing how various meteorological factors can enhance or diminish sighting opportunities. Participants can share real-time reports on local weather conditions, helping others plan their observations more effectively. This collaborative approach not only fosters a sense of community but also underscores the importance of understanding the environment when attempting to witness unidentified aerial phenomena.

Finally, engaging with UFO communities through online forums and social media groups allows for the documentation and reporting of sightings in an organized manner. Many platforms encourage users to share their findings with specific details, creating a repository of information that can be analyzed scientifically. This collective effort contributes to a broader understanding of UFO phenomena, enriching discussions about the psychology of UFO beliefs and sightings. By participating in these online spaces, individuals not only enhance their own knowledge but also contribute to a growing body of research that seeks to unveil the mysteries of the universe.

Local UFO Clubs and Organizations

Local UFO clubs and organizations serve as vital hubs for enthusiasts, researchers, and anyone intrigued by the mysteries of unidentified flying objects. These clubs often attract a diverse membership, including UFO believers, individuals who have encountered the supernatural, and those simply curious about the phenomena. By fostering a sense of community, local organizations provide a platform for sharing experiences, exchanging knowledge, and collaborating on research initiatives. Members can engage in discussions about their sightings, theories, and the latest developments in UFO research, creating an environment where curiosity and inquiry thrive.

One of the primary benefits of joining a local UFO club is access to educational resources and expert-led discussions. Many organizations host guest speakers, including researchers, authors, and even former military personnel who have had firsthand experiences with UFOs. These events can provide invaluable insights into how to spot UFOs, as well as the techniques and technologies available for night sky observation. Members often share tips on using apps and telescopes, which can enhance one's ability to identify potential UFO sightings amidst common misidentifications such as aircraft or satellites.

Moreover, local UFO clubs frequently organize group outings for skywatching. These events not only allow members to observe the night sky together but also facilitate real-time discussions about weather conditions that can impact UFO visibility. By understanding how atmospheric phenomena influence sightings, enthusiasts can better prepare for their observations and improve their chances of witnessing unidentified objects. Additionally, clubs often provide training sessions on advanced photography techniques, equipping members with the skills needed to capture clear images of potential UFOs for documentation and analysis.

Engaging with local UFO communities can significantly enhance one's experience in documenting and reporting sightings. Many organizations have established protocols for reporting incidents, ensuring that information is gathered systematically and can be shared with broader research efforts. This structured approach not only validates personal experiences but also contributes to a larger dataset that can be analyzed scientifically. Members are encouraged to think critically about their experiences, fostering a scientific mindset that is essential for understanding the complexities of UFO sightings.

Lastly, the psychology of UFO beliefs is an intriguing area of exploration within these communities. Local clubs often facilitate discussions about the cultural and psychological factors influencing how individuals perceive and interpret their experiences with the unknown. Understanding these dynamics can help enthusiasts contextualize their own beliefs and sightings, encouraging a more nuanced approach to the subject. By connecting with like-minded individuals in a supportive environment, members can deepen their understanding of the phenomena while contributing to a collective pursuit of knowledge and truth in the mysterious world of UFOs.

Notable Conferences and Events

Notable conferences and events dedicated to UFO phenomena play a crucial role in fostering community, sharing knowledge, and advancing the dialogue surrounding unidentified aerial phenomena. These gatherings offer enthusiasts, researchers, and believers alike a platform to exchange experiences, discuss findings, and explore the latest theories and technologies related to UFO sightings. From local meet-ups to international conventions, these events encompass a wide range of topics relevant to understanding the unknown and provide invaluable networking opportunities among individuals who share a passion for the supernatural.

One of the most significant events on the UFO calendar is the International UFO Congress. This annual gathering attracts researchers, experiencers, and enthusiasts from across the globe to share their insights and investigate the latest evidence in UFO research. With a lineup that includes leading experts, panel discussions, and workshops, attendees have the chance to delve into various subjects, such as advanced photography techniques for capturing UFOs, the role of weather conditions in visibility, and the psychological aspects of UFO beliefs. The congress not only facilitates learning but also encourages collaboration among participants, helping to bridge the gap between amateur enthusiasts and seasoned researchers.

Another notable event is the MUFON Symposium, organized by the Mutual UFO Network, which emphasizes the importance of scientific inquiry into UFO sightings. This symposium showcases presentations from a diverse array of speakers—including scientists, military personnel, and experiencers—who discuss their findings and experiences. Topics often include the analysis of UFO sightings through a scientific lens, effective documentation and reporting methods for sightings, and how to engage with local UFO communities. By fostering a scientific approach to the study of UFOs, the symposium promotes an understanding that transcends anecdotal evidence and encourages rigorous investigation.

The Contact in the Desert event is also a highlight for those interested in the intersection of UFO phenomena and spirituality. This festival features a variety of speakers who explore the broader implications of UFO sightings, including discussions on extraterrestrial contact, consciousness, and the nature of reality. Workshops on night sky observation techniques and the use of technology for UFO spotting are often included, allowing attendees to enhance their skills while engaging with like-minded individuals. The event serves not only as an educational experience but also as a community-building opportunity for those who have experienced the supernatural or seek to understand it better.

Additionally, smaller, grassroots gatherings provide vital spaces for local communities to connect and share their experiences. These events often focus on specific regions or themes, allowing participants to discuss local sightings, misidentifications, and effectively analyze reports. They can include workshops on using apps and telescopes for UFO spotting, as well as sessions dedicated to sharing personal stories of encounters. These intimate gatherings create a welcoming environment for newcomers and seasoned enthusiasts alike, fostering a sense of belonging and shared purpose in the quest to unveil the mysteries of the skies. By participating in these notable conferences and events, individuals can deepen their understanding of UFO phenomena and contribute to the collective knowledge of this intriguing field.

Chapter 9

Documenting and Reporting UFO Sightings Effectively

Importance of Accurate Documentation

Accurate documentation is fundamental to the study and understanding of UFO phenomena. For those who identify as UFO believers or have experienced the supernatural, the act of thoroughly recording sightings serves multiple purposes. It not only preserves the details of an event for future analysis but also enhances the credibility of the report. In an area often met with skepticism, precise documentation can transcend anecdotal evidence, providing a solid foundation for further investigation and discourse within the community. An accurate record can help differentiate genuine experiences from common misidentifications, such as aircraft or satellites, which is crucial for fostering a clearer understanding of what is truly observed in the night sky.

One of the most critical elements of effective documentation is the inclusion of specific details surrounding the sighting. This includes the date, time, and location of the event, as well as environmental conditions such as weather, visibility, and any notable atmospheric phenomena. For enthusiasts focusing on night sky observation techniques, noting the position of celestial bodies can aid in distinguishing between UFOs and familiar objects. The role of technology, like mobile apps and telescopes, can be invaluable in this regard, as they help to pinpoint the locations of traditional aerial phenomena, thereby sharpening one's observational skills and improving the accuracy of future reports.

Moreover, advanced photography tips play a pivotal role in documenting UFO sightings. Capturing high-quality images or videos can provide visual evidence that bolsters personal accounts and adds weight to the collected data. Techniques such as using manual settings on cameras, employing stabilizers, and understanding the impact of lighting conditions can enhance the quality of documentation. With many enthusiasts often relying on smartphone cameras, it is essential to understand their limitations and how to work around them, ensuring that when a potential sighting occurs, the documentation captures as much detail as possible.

Engaging with UFO communities and resources is another vital aspect of accurate documentation. By sharing experiences with like-minded individuals, one can gain insights into effective reporting techniques and learn from the collective knowledge of the community. This engagement encourages the sharing of documented findings, which can lead to collaborative analysis and a deeper understanding of the phenomena. Moreover, when individuals report sightings to established organizations, they contribute to a broader database that can help researchers identify patterns and trends, ultimately advancing the scientific approach to UFO studies.

Finally, the psychological aspects of UFO beliefs and sightings cannot be overlooked. Accurate documentation can serve as a tool for individuals to process their experiences and challenge their perceptions of reality. By recording details meticulously, believers can navigate the often tumultuous terrain of skepticism and belief. This process not only aids in personal understanding but also contributes significantly to the collective narrative surrounding UFO phenomena, fostering a culture of inquiry and evidence-based discussion. In this way, accurate documentation becomes not just an exercise in detail, but a vital component of the journey toward understanding the unknown.

How to Create a UFO Sightings Report

Creating a UFO sightings report is a crucial step for anyone interested in documenting their encounters with the unexplained. Whether you are a seasoned UFO believer or a newcomer to the phenomenon, a well-structured report enhances credibility and aids in the overall understanding of such experiences. The first step in drafting a UFO sightings report involves detailing the specifics of the event. This includes the exact date and time of the sighting, the location (preferably with coordinates), and a descriptive account of what was observed. Observers should note the shape, color, and size of the object, as well as any movements or unusual behaviors that stood out during the encounter.

Next, it is essential to provide context about the environment at the time of the sighting. This includes documenting weather conditions, such as cloud cover, visibility, and any atmospheric phenomena that may have influenced the sighting. For example, overcast skies or fog can affect how well an object can be seen. Additionally, noting any nearby structures, landmarks, or other objects can help in establishing a frame of reference. Understanding the surroundings can also assist in addressing potential misidentifications, such as distinguishing between aircraft, satellites, or natural phenomena like meteors.

Using technology can significantly enhance the quality of your UFO sightings report. If you are equipped with a smartphone or a camera, capturing images or videos of the sighting can provide invaluable evidence. When documenting a sighting, ensure that you include the settings used, such as exposure time and any filters applied. Many apps are available that can aid in tracking celestial bodies and identifying aircraft, which can be referenced to help validate or debunk your sighting. Remember that thorough documentation can make your report more compelling and informative, especially when engaging with UFO communities or researchers.

Engaging with the wider UFO community is another vital aspect of reporting a sighting. Sharing your experience through forums, local meetups, or dedicated websites allows others to provide insights or corroborate similar experiences. Many organizations have established guidelines for reporting sightings, which can add an additional layer of credibility to your report. When submitting your findings, be clear and concise, and provide all relevant details to ensure your report is useful to those who may analyze or investigate it further.

Finally, a comprehensive UFO sightings report should include a personal reflection on the experience. This could encompass your emotional response, any changes in your perspective on UFOs, or how the encounter has influenced your beliefs. Such reflections not only contribute to the narrative but also may aid in understanding the psychology behind UFO sightings. Documenting personal insights can be just as significant as the factual data collected, offering a holistic view of the experience that resonates with fellow enthusiasts and researchers alike. By following these guidelines, you can create an informative and impactful UFO sightings report that contributes to the ongoing exploration of the unknown.

Sharing Your Sightings with the Community

Sharing your sightings with the community is a vital step in the ongoing exploration and understanding of UFO phenomena. When you encounter an unidentified object in the sky, documenting and sharing your experience not only contributes to the collective knowledge but also helps foster a sense of camaraderie among fellow enthusiasts. Engaging with others who share your passion can provide valuable insights, support, and a platform for discussing theories and experiences.

To effectively share your sightings, begin by documenting every detail of the event. This includes the date, time, location, weather conditions, and any unusual behaviors exhibited by the object. Photographic evidence, when available, can significantly enhance your report, so consider employing advanced photography techniques. Utilizing a tripod, adjusting camera settings for low light, and even using specialized apps can help capture clearer images of potential UFOs. Remember that an accurate and thorough report can serve as an essential resource for both amateur and professional researchers in the field.

Once you have gathered your documentation, consider engaging with local and online UFO communities. Forums, social media groups, and dedicated websites are excellent venues for sharing your findings. When you post your sighting, be prepared for questions and discussions that may follow. This interaction can lead to a deeper understanding of the phenomena, as members may provide their interpretations or share similar experiences. Additionally, these communities often have access to resources, such as research databases and expert contacts, that can enhance your understanding of what you witnessed.

It is also important to be aware of common misidentifications when sharing your sightings. Many UFO reports are later identified as natural phenomena, aircraft, or satellites. By familiarizing yourself with these common misidentifications, you can provide a more comprehensive account of your experience and help others refine their observational skills. Engaging in discussions about these misidentifications can also demystify UFO sightings, allowing for a more scientific approach to analysis, which is essential for building credibility within the community.

Lastly, consider the role of technology in sharing your sightings. Utilizing apps designed for sky observation can help you and fellow enthusiasts track celestial events, enhance your spotting capabilities, and facilitate discussions on what you observe. By harnessing these tools and sharing your experiences, you contribute to a growing body of knowledge that encourages critical thinking and informed discourse. Embracing the collective effort of documenting, analyzing, and discussing sightings will not only enrich your understanding but also strengthen the community's commitment to uncovering the mysteries of the unknown.

Chapter 10

Analyzing UFO Sightings: A Scientific Approach

The Importance of Objectivity

In the realm of UFO sightings and supernatural experiences, the pursuit of objectivity is paramount. Objectivity allows individuals to approach phenomena without bias, fostering a more accurate understanding of what they witness. For UFO believers and enthusiasts, this means differentiating between genuine unidentified flying objects and common misidentifications such as aircraft, satellites, or atmospheric phenomena. By cultivating an objective mindset, observers can enhance their observational skills, leading to more credible reports and a deeper understanding of the unknown.

One of the primary challenges in the study of UFOs is the heavy influence of personal beliefs and emotions on perceptions. When individuals approach sightings with preconceived notions, they risk overlooking critical details or misinterpreting the evidence. This is particularly crucial for those engaged in night sky observation techniques, where distractions and biases can cloud judgment. By emphasizing objectivity, enthusiasts can develop a systematic approach to spotting UFOs, focusing on concrete evidence and observable characteristics rather than assumptions based on anecdotal experiences.

Utilizing technology can significantly aid in maintaining objectivity in UFO spotting. Various apps and tools designed for stargazing, alongside advanced photography techniques, enable enthusiasts to capture and analyze sightings with precision. Objectivity in this context means using these resources to validate observations rather than to sensationalize them. For example, employing an app that identifies known celestial bodies can help differentiate between a satellite and a potential UFO, thereby refining the observer's understanding and reducing the likelihood of misidentification.

Weather conditions also play a critical role in UFO visibility, making it essential for observers to remain objective when considering how external factors influence sightings. Atmospheric phenomena, such as cloud cover or humidity, can obscure or distort what is seen in the sky. By recognizing the impact of weather on visibility, individuals can approach their observations with a more discerning eye, systematically ruling out environmental variables that could explain their experiences. This analytical approach not only enhances personal credibility but also contributes to a collective understanding of UFO phenomena.

Finally, engaging with UFO communities and resources can further promote objectivity among enthusiasts. Collaborative discussions and shared experiences can provide valuable insights, helping to refine individuals' approaches to documenting and reporting sightings. By fostering a culture that prioritizes objective analysis over sensationalism, the UFO community can work together to advance scientific understanding of these mysterious occurrences. Ultimately, embracing objectivity is not just about improving personal observation skills; it is about contributing to a broader discourse that seeks to unveil the truth behind the unknown.

Tools for Analysis and Comparison

In the quest to understand the phenomenon of UFOs, a structured approach to analysis and comparison is essential for both enthusiasts and serious researchers. This subchapter will outline the various tools available to aid in the investigation of unidentified flying objects, ensuring that believers and skeptics alike can engage in informed discussions and analyses. By employing the right resources, individuals can enhance their observational skills, differentiate between genuine sightings and common misidentifications, and contribute meaningfully to the discourse surrounding the unexplained.

To begin, it's important to equip oneself with observational tools that enhance the ability to spot UFOs. Night sky observation techniques involve a combination of the naked eye and technology. Binoculars and telescopes can significantly improve visibility, allowing observers to discern details that are often missed. Additionally, there are numerous smartphone apps designed to assist in identifying celestial objects. These applications can provide real-time data on satellites, aircraft, and other aerial phenomena, helping to clarify whether a sighting is a UFO or a common object. By leveraging these tools, enthusiasts can refine their skills and gain a deeper understanding of what they observe.

Weather conditions play a pivotal role in the visibility of UFOs, making meteorological tools vital in this analysis. Understanding how factors such as cloud cover, humidity, and atmospheric disturbances affect visibility can provide crucial context for sightings. Observers should familiarize themselves with local weather patterns and utilize online resources to predict optimal viewing conditions. By correlating sightings with specific weather occurrences, researchers can begin to identify trends that may reveal insights into when and where UFOs are more likely to be seen.

For those looking to document their sightings effectively, advanced photography techniques are indispensable. High-quality cameras, tripods, and proper settings can capture fleeting moments that the naked eye might miss. Effective use of zoom, exposure, and focus can transform a blurry image into a valuable piece of evidence. Moreover, understanding how to analyze these photographs scientifically can help distinguish between genuine anomalies and artifacts caused by camera limitations or environmental factors. This analytical approach not only strengthens personal documentation but also contributes to a larger body of evidence within UFO research.

Engaging with UFO communities and resources can further enhance an individual's analytical capabilities. Online forums, social media groups, and local meetups provide platforms for sharing experiences and insights, fostering collaboration among UFO enthusiasts. Participating in these communities can yield access to collective knowledge, including best practices for reporting sightings and analyzing data. By documenting and sharing findings, individuals contribute to a growing database of information that can be scrutinized and compared, ultimately enriching the understanding of UFO phenomena. This collaborative effort is essential in demystifying the unknown and advancing the discourse surrounding UFOs.

Case Studies of Notable Sightings

In the realm of UFO sightings, numerous cases have captured the imagination of both enthusiasts and skeptics alike. This subchapter delves into notable sightings that have not only sparked curiosity but also provided valuable insights into understanding unidentified aerial phenomena. Each of these cases offers a unique perspective on the complexities of UFO observation, the potential for misidentification, and the broader implications for society's understanding of the unknown.

One of the most famous cases is the 1973 Pascagoula incident, where two men reported being abducted by an alien craft while fishing in Mississippi. Their detailed accounts of the encounter, characterized by a glowing light and humanoid figures, prompted investigations by law enforcement and researchers alike. This case highlights the importance of eyewitness testimony and the psychological factors that can influence perceptions of reality. The Pascagoula incident not only ignited public interest but also underscored the need for rigorous documentation and reporting of such events to separate fact from fiction.

Another significant sighting occurred in 1997 during the Phoenix Lights phenomenon, where thousands of witnesses observed a series of V-shaped lights moving silently across the Arizona sky. The event garnered extensive media coverage and has been analyzed from various angles, including meteorological conditions and potential military aircraft operations. This sighting exemplifies how environmental factors, such as atmospheric conditions and light reflection, can affect visibility and lead to misinterpretations of what is seen in the night sky. It emphasizes the necessity for UFO observers to consider these variables while documenting their experiences.

The 2006 O'Hare International Airport sighting further complicates the narrative around UFOs. Ground crew and pilots reported seeing a disc-shaped object hovering above the airport before it abruptly ascended through the cloud cover. This incident has been subjected to scrutiny by both aviation experts and UFO researchers, highlighting the intersection of technology and human observation. It serves as a reminder of the critical role that modern technology, including radar and photographic equipment, plays in corroborating or refuting claims of UFO sightings. Enthusiasts can benefit from understanding these technologies to enhance their own observation efforts in the field.

Lastly, the 2017 revelations of Navy pilots encountering unidentified aerial phenomena during training exercises have reignited discussions about government disclosure and the legitimacy of UFO sightings. The encounters, captured on advanced surveillance systems, provided compelling visual evidence that has since sparked a wave of public interest and calls for transparency. Analyzing these cases through a scientific lens is essential for both believers and skeptics alike, as it encourages a thoughtful examination of the evidence and fosters a sense of community among those engaged in the pursuit of understanding the unknown.

These case studies not only illustrate the diversity of UFO sightings but also emphasize the importance of rigorous analysis, community engagement, and the application of technology in the quest for truth. By examining these notable incidents, enthusiasts can cultivate a more informed approach to spotting UFOs, documenting experiences, and contributing to the ongoing dialogue surrounding the mysteries of the night sky.

Chapter 11

The Psychology of UFO Beliefs and Sightings

Understanding the Human Mind and Perception

Understanding the human mind and perception is crucial for anyone delving into the world of UFOs and supernatural experiences. Our perception shapes how we interpret the world around us, including the night sky and the phenomena within it. The human mind is an intricate system that processes sensory information, but it is also subject to biases and illusions that can distort our understanding. This chapter will explore how cognitive processes influence our observations and interpretations of unidentified flying objects, highlighting the importance of critical thinking in the pursuit of truth.

Perception is not merely a passive reception of information; it is an active process influenced by our beliefs, expectations, and experiences. Individuals interested in spotting UFOs may find themselves predisposed to interpret ambiguous stimuli—such as lights in the sky—as extraterrestrial craft. This phenomenon, known as pareidolia, occurs when the brain perceives familiar patterns in random stimuli, leading to misinterpretations. Understanding this cognitive tendency can help enthusiasts approach sightings with a more analytical mindset, enabling them to differentiate between genuine anomalies and commonplace explanations like aircraft or satellites.

Moreover, the context in which an observer finds themselves can heavily influence their perceptions. Environmental factors, such as lighting conditions, weather patterns, and even emotional states, can alter what is seen in the night sky. For instance, atmospheric conditions like humidity and temperature can create optical illusions, causing objects to appear distorted or misplaced. UFO seekers should consider these variables when documenting sightings, as they can significantly impact visibility and clarity. By recognizing the interplay between mind and environment, observers can enhance their night sky observation techniques, improving their chances of identifying true UFO phenomena.

Technology has become an invaluable tool for UFO enthusiasts, offering advanced methods for capturing and analyzing potential sightings. Apps designed for sky observation can help users identify stars, planets, and satellites, minimizing the likelihood of misidentification. Additionally, using telescopes equipped with high-resolution cameras can facilitate the documentation of unusual objects, allowing for detailed analysis later. However, even with technology at their disposal, individuals must remain vigilant about the psychological aspects of perception. A scientific approach to analyzing sightings is essential, as it mitigates the risk of confirmation bias, where observers disproportionately favor evidence that supports their beliefs.

Lastly, the psychology behind UFO beliefs and sightings is a fascinating area of study that reveals much about human nature. Many who report encounters do so with a genuine sense of wonder and curiosity, seeking to understand phenomena that challenge the boundaries of known science. Engaging with UFO communities can provide support and resources for those navigating these experiences, fostering a sense of belonging among fellow enthusiasts. Ultimately, by combining an understanding of human perception with technological advancements and community engagement, individuals can enhance their UFO spotting endeavors and contribute to a more informed discourse about the unknown.

The Role of Cognitive Bias in Sightings

Cognitive bias significantly influences how individuals interpret and report their experiences with unidentified flying objects (UFOs). This phenomenon occurs when people's perceptions and memories are shaped by their beliefs, expectations, and prior knowledge, rather than objective reality. For UFO believers and supernatural enthusiasts, cognitive biases can enhance the thrill of a sighting but may also lead to misconceptions and misinterpretations. Understanding these biases is essential for anyone interested in accurately assessing sightings and engaging with UFO communities.

One common type of cognitive bias is confirmation bias, where individuals tend to favor information that confirms their pre-existing beliefs. For example, a person who strongly believes in extraterrestrial life may interpret ambiguous lights in the sky as UFOs, disregarding other possible explanations. This tendency can skew personal accounts of sightings, as witnesses often emphasize details that support their beliefs while downplaying or ignoring contradictory evidence. Recognizing this bias can help enthusiasts approach sightings with a more critical mindset, enhancing their ability to distinguish genuine phenomena from misidentified objects.

Another relevant cognitive bias is the availability heuristic, which leads individuals to overestimate the likelihood of events based on how easily examples come to mind. If someone has recently encountered numerous reports of UFO sightings or watched films depicting alien encounters, they may be more likely to perceive ordinary occurrences—like a plane or a satellite—as UFOs. This bias can be particularly pronounced during times of heightened media coverage about UFOs or related phenomena. By understanding how this heuristic operates, observers can develop strategies for more accurate identification, taking into account the broader context of their experiences.

The influence of social and cultural factors cannot be understated when examining cognitive biases in UFO sightings. Groupthink, for instance, can occur within UFO communities, where individuals conform to the dominant beliefs or experiences shared by others. This phenomenon can lead to the amplification of certain sightings and the dismissal of alternative explanations, creating echo chambers that reinforce existing biases. Engaging with diverse perspectives and seeking out scientific resources can help mitigate these effects and promote a more balanced understanding of potential UFO phenomena.

Lastly, the psychology behind UFO beliefs and sightings is interconnected with the very nature of human perception. Our brains are wired to seek patterns and make sense of the unknown. This innate tendency can lead to pareidolia, where individuals perceive familiar shapes, such as faces or objects, in random stimuli. Such experiences can fuel the belief in UFOs, as people may attribute meaning to fleeting glimpses of light or movement in the sky. By acknowledging these psychological dynamics, UFO enthusiasts can cultivate a more nuanced approach to sightings, combining personal experiences with a scientific understanding of perception and cognition, ultimately enhancing their night sky

observation techniques and documentation efforts.

Cultural Influences on UFO Beliefs

Cultural influences play a significant role in shaping beliefs and narratives surrounding UFO phenomena. Across different societies, the interpretation of unidentified flying objects is often filtered through local mythology, historical events, and cultural symbolism. In many cases, UFO sightings are intertwined with the broader context of societal beliefs about the unknown, the supernatural, and the possibility of extraterrestrial life. These cultural narratives can enhance or diminish the perceived credibility of UFO encounters, as well as influence how individuals interpret their experiences with the unknown.

In Western cultures, for instance, the mid-20th-century UFO wave was closely linked to the rise of science fiction, with films and literature portraying extraterrestrial beings as both benevolent and malevolent. This duality has shaped public perception, where sightings are often colored by the prevailing narratives in popular media. The influence of iconic films like "Close Encounters of the Third Kind" and "The X-Files" cannot be overstated; they have created archetypes that inform how people interpret their sightings. The interplay between fiction and reality often leads to an expectation of what a UFO should look like, which can impact the reporting and documentation of actual sightings.

Conversely, in non-Western cultures, UFO beliefs may intersect with ancient mythologies and spiritual practices. For example, some indigenous cultures view UFOs as manifestations of ancestral spirits or extraterrestrial visits as part of a larger cosmic order.

These interpretations can provide a sense of reassurance and connection to the universe, contrasting sharply with the often fearful narratives found in Western accounts. Understanding these cultural variations is crucial for anyone interested in UFO phenomena, as they highlight the diverse ways in which humans seek to make sense of the unknown.

Moreover, the role of technology in modern UFO spotting is also influenced by cultural factors. The proliferation of smartphones and dedicated apps for tracking celestial events has democratized access to the night sky, enabling enthusiasts to engage with their environment actively. However, the effectiveness of these technologies can be influenced by cultural attitudes towards science and skepticism. In cultures that embrace technological advancements, there is often a more rigorous approach to investigating sightings, using documentation and analysis to validate experiences. In contrast, communities that view technology with suspicion may rely on traditional storytelling and oral histories, which can lead to vastly different methods of engagement and understanding.

Finally, the psychology of UFO beliefs often reflects broader societal anxieties and hopes. In times of social upheaval or uncertainty, people may gravitate towards UFO narratives as a means of coping. The quest for disclosure—an understanding of what lies beyond our planet—often mirrors deeper existential questions about humanity's place in the universe. This psychological dimension intertwines with cultural influences, creating a complex web of belief systems that inform how individuals perceive and react to UFO sightings. By recognizing these cultural and psychological layers, UFO enthusiasts can gain a more nuanced understanding of their experiences, ultimately enriching their exploration of the unknown.

Chapter 12

Conclusion and Future Directions

The Evolving Landscape of UFO Research

The landscape of UFO research has undergone significant transformation over the decades, shifting from fringe theories and sensational tabloid headlines to a more structured and scientific inquiry. This evolution is marked by a growing acknowledgment of the phenomenon among mainstream scientists, government agencies, and the public. In recent years, the declassification of military reports and the establishment of dedicated task forces have fueled interest and legitimacy in UFO research, allowing enthusiasts and skeptics alike to engage with the topic more critically and constructively. The shift from mere speculation to investigative rigor marks a pivotal moment for those who seek to understand the unknown.

The advent of technology has played a crucial role in this evolving landscape. With the proliferation of smartphones and high-quality cameras, individuals are now equipped to capture evidence of potential UFO sightings like never before. Various apps designed for night sky observation enhance the ability of enthusiasts to identify celestial bodies, track satellites, and distinguish between aircraft and unidentified phenomena. This technological advancement empowers ordinary people to contribute to the body of evidence surrounding UFO sightings, while also improving the overall understanding of what constitutes a legitimate observation versus a common misidentification.

Research methodologies have also adapted to incorporate scientific approaches to analyzing UFO sightings. This includes investigating the psychological factors that drive belief in UFOs and the mechanisms behind eyewitness testimony. By employing techniques from fields such as psychology and sociology, researchers aim to separate genuine encounters from those influenced by cognitive biases or societal pressures. Furthermore, documenting and reporting sightings has become more structured, with platforms and resources available for UFO enthusiasts to share their experiences in a systematic manner that encourages thorough investigation and analysis.

Weather conditions, often overlooked, are another crucial element in UFO visibility and research. Atmospheric phenomena such as clouds, temperature inversions, and light refraction can significantly impact how objects are perceived in the night sky. Understanding these factors not only aids in accurately assessing sightings but also enhances the skills of observers. By integrating knowledge of meteorological conditions with night sky observation techniques, enthusiasts can refine their ability to distinguish between genuine unidentified objects and those that can be readily explained through natural occurrences.

Engagement with UFO communities has also flourished, providing rich resources for enthusiasts looking to deepen their understanding of the phenomenon. Online forums, local meetups, and organized conferences allow individuals to share experiences, analyze sightings collaboratively, and learn from experts in the field. This communal approach fosters an environment of support and knowledge sharing, which is vital for both newcomers and seasoned researchers. As the landscape of UFO research continues to evolve, it becomes increasingly clear that a combination of technology, community engagement, and scientific inquiry will shape the future of how we understand and interact with the unknown.

Encouraging Critical Thinking and Open Dialogue

Encouraging critical thinking and open dialogue is essential in the exploration of UFO phenomena, especially for those who are passionate believers or have had supernatural experiences. The field of ufology often attracts a diverse range of opinions and interpretations, which can lead to misunderstandings and polarization among enthusiasts. By fostering a culture of inquiry and respect, we can deepen our understanding of the unknown while also enhancing our collective experience as observers of the night sky.

Critical thinking involves the ability to analyze and evaluate evidence systematically. When it comes to UFO sightings, this means scrutinizing every detail—from the location and time of the sighting to the conditions under which it occurred. For instance, understanding common misidentifications, such as aircraft, satellites, or atmospheric phenomena, is crucial. By encouraging enthusiasts to engage with these concepts, we empower them to distinguish genuine anomalies from ordinary occurrences. This analytical approach not only aids in personal understanding but also equips individuals to share their insights more effectively within UFO communities.

Open dialogue is equally important. Engaging in discussions with others who share a passion for the supernatural can lead to new perspectives and insights. UFO enthusiasts should feel encouraged to share their experiences and theories without fear of ridicule. Creating spaces for respectful conversation allows us to connect with diverse viewpoints, which can enhance our understanding of the phenomena we observe. Whether through online forums, local meetups, or organized events, fostering open communication helps build a supportive community where individuals feel comfortable exploring their beliefs and experiences.

Using technology can also play a significant role in promoting critical thinking and dialogue. There are numerous apps and tools available that assist in UFO spotting and documentation. These resources can help enthusiasts verify sightings by comparing real-time data with reported observations. By utilizing technology effectively, individuals can not only improve their observational skills but also contribute to a larger database of sightings, enriching the discourse surrounding UFO phenomena. Encouraging the use of advanced photography techniques and weather condition analysis can further deepen our understanding and documentation of these mysterious events.

Finally, analyzing UFO sightings through a scientific lens promotes a balanced approach to the subject. By applying the principles of scientific inquiry, enthusiasts can assess the validity of sightings, which often leads to more credible discussions. This approach also helps to demystify the psychology behind UFO beliefs, allowing individuals to explore their motivations and experiences without bias. Ultimately, by encouraging critical thinking and open dialogue, we can cultivate a more informed, respectful, and engaging community of UFO enthusiasts, dedicated to unraveling the mysteries of the unknown.

Resources for Continued Learning and Exploration

In the quest for understanding UFO phenomena and enhancing your observational skills, a wealth of resources is available to deepen your knowledge and refine your techniques. Whether you're a seasoned enthusiast or newly captivated by the mysteries of the night sky, these resources can help elevate your experience and understanding of unidentified flying objects. From literature to technological tools, the following sections outline various avenues for continued learning and exploration.

Books and documentaries serve as foundational resources for anyone interested in UFOs. Numerous authors have contributed to the field, presenting a range of perspectives from skeptical analyses to enthusiastic accounts of personal experiences. Notable titles often cover topics such as historical sightings, government disclosures, and the psychology behind UFO beliefs. Documentaries can provide visual evidence and expert interviews, enhancing your understanding of complex concepts. Engaging with this literature can broaden your viewpoint and inspire deeper inquiry into the subject.

Technology plays a crucial role in modern UFO spotting, offering enthusiasts innovative ways to document and analyze sightings. Smartphone applications designed for stargazing not only help identify celestial objects but also allow users to log their UFO encounters in real-time. Additionally, advanced telescopes equipped with digital tracking systems can aid in capturing elusive objects that might otherwise go unnoticed. Understanding how to effectively use these tools can significantly enhance your observational capabilities, making it easier to distinguish between genuine UFOs and common misidentifications such as aircraft or satellites.

Weather conditions are a vital, yet often overlooked, aspect of UFO visibility. To maximize your chances of spotting unidentified objects, it's essential to understand how various weather phenomena can affect visibility. Resources that focus on meteorology can provide insights into cloud cover, atmospheric clarity, and light pollution—all of which play significant roles in your observational success. By learning to analyze weather patterns, you can choose optimal times and locations for your skywatching endeavors, increasing the likelihood of encountering something extraordinary.

Engaging with UFO communities can also enrich your journey into the unknown. Online forums and local meetups provide platforms for sharing experiences, discussing theories, and collaborating on investigations. These communities often have access to valuable resources, including workshops and guest speakers who can offer expert knowledge. Additionally, many enthusiasts maintain databases for documenting sightings, which can serve as a rich repository for analysis. By actively participating in these networks, you can not only share your experiences but also gain insights from others who share your passion for the supernatural.

Lastly, honing your skills in documenting and analyzing UFO sightings is essential for contributing meaningfully to the field. Resources that focus on scientific approaches to UFO analysis can help you develop a critical eye and systematic methodology for evaluating encounters. Understanding the psychological aspects of UFO beliefs can also provide context for your observations and the experiences of others. By combining technical knowledge, community engagement, and scientific inquiry, you can cultivate a well-rounded perspective that contributes to the broader conversation surrounding UFOs and the mysteries they embody.

www.ingramcontent.com/pod-product-compliance
Lightning Source LLC
Chambersburg PA
CBHW021804130726
47987CB00008B/3008